POCKET
EVOLUTION
THINGS YOU SHOULD KNOW
(QUESTIONS AND ANSWERS)

Rumi Michael Leigh

Introduction

I would like to thank and congratulate you for downloading this book, *"Pocket Evolution, things you should know (questions and answers)"* series.

This book will give you a good general knowledge about the essentials of Evolution.

Thanks again for downloading this book, I hope you enjoy it!

Chapter 1: Questions

1. Define evolution.
2. Name a factor that proves evolution.
3. Explain evolutionism.
4. Who found the concept of evolutionism?
5. What is adaptation?
6. What is adaptive radiation?
7. What are species?
8. What is coevolution?
9. What is variation?
10. What is morphology?

Chapter 1: Answers

1. Evolution is the gradual modification of the genetic composition of a population.
2. The presence of fossils.
3. This is the concept that species are gradually drifting from each other by natural selection.
4. Charles Darwin.
5. Adaptation is the modification of the structures favourable to the survival of an organism.
6. Adaptive radiation is the rapid immergence of many species from a common ancestor.
7. These are similar organisms that can reproduce together and have offsprings that can also reproduce.
8. This is the reciprocal changes and evolution between two species.
9. This is the difference in DNA.
10. This is the form, shape, and structure of living organisms.

Chapter 2: Questions

1. How many species have been identified?
2. Half of the identified species are?
3. What is taxonomy or systematics?
4. Why is the old classification of species false?
5. What is the main difference between eukaryotes and prokaryotes?
6. According to biochemistry, molecular biology and microbiology, what are the ancestors of all living things?
7. What are the oldest known fossils?
8. What is the great success of prokaryotes?
9. What is uniformitarianism?
10. What is the orthogenetic theory?

Chapter 2: Answers

1. Over 1.5 million.
2. Insects.
3. Taxonomy or systematics is the classification of species.
4. It is false because it assumes that some species are more important than others. And normally, there is no order of importance. Each specie has different functions and is also as important.
5. Eukaryotes have a true nucleus but prokaryotes do not have a true nucleus.
6. Prokaryotes.
7. Prokaryotes.
8. They can reproduce quickly in a favourable environment.
9. This is a concept that the earth's surface was gradually shaped by forces such as erosion, etc.
10. This is the theory that states that natural selection is not required for evolution.

Chapter 3: Questions

1. What is natural selection?
2. Name 2 conditions that favour natural selection.
3. What are the 3 types of natural selection?
4. What is disruptive selection?
5. What is stabilizing selection?
6. What is directional selection?
7. What is artificial selection?
8. Where are artificial selections used?
9. What is emigration?
10. What is immigration?

Chapter 3: Answers

1. This consists of the traits that make an organism best suited to an environment.
2. A population that is being predated. A population where there is not enough resource.
3. Disruptive selection, directional selection, and stabilizing selection.
4. It is a selection that eliminates intermediate types, so it favours both extremes.
5. It is the selection that favours the characteristics of the intermediate types.
6. It is the selection that favours one extreme direction.
7. This is the selective breeding of plants and animals chosen by humans.
8. In animal domestication and agriculture.
9. This is when an organism moves out into another location.
10. This is when an organism moves into a new location.

Chapter 4: Questions

1. Explain fixism.
2. Explain the concept of the inheritance of acquired characters.
3. Who found the concept of the inheritance of acquired characters?
4. What example did Lamarck give for acquired characters?
5. Explain spontaneous generation.
6. Explain totalitarian heredity.
7. Why was Stanley Miller's test a half-fail?
8. What were the substances of life in the Stanley Miller test?
9. Explain the SOMA-GERMIN concept.
10. Who found the concept of SOMA-GERMIN?
11. Who found the concept in response to the spontaneous generation that life always comes from life?
12. Who found the particular inheritance of genes?

Chapter 4: Answers

1. This is the concept that species are created from the beginning by God.
2. It is the concept that acquired characters can be passed from one generation to the next.
3. Lamarck.
4. The stretching of giraffe.
5. This is the concept that explains the spontaneous occurrence of micro-organisms in certain nutrient medium.
6. This is the concept that speaks of homunculus, where there is not much female participation.
7. It was a half-fail because there was no appearance of life in the test tubes but substances of life.
8. Amino acids, carbohydrates, lipids, etc.
9. It is the concept that only germ cells transmit hereditary traits and that acquired traits cannot be transmitted.
10. Weismann.
11. Louis Pasteur.
12. Mendel.

Chapter 5: Questions

1. What is a mineral?
2. What is a rock?
3. Name the 3 types of rocks.
4. How are sedimentary rocks formed?
5. How are magmatic rocks formed?
6. How are metamorphic rocks formed?
7. What is stratigraphy?
8. What is palaeontology?
9. What is a fossil?
10. What is half-life?

Chapter 5: Answers

1. A mineral is a homogeneous crystalline solid with a precise chemical formula.
2. A rock is a mixture of several minerals.
3. Sedimentary rocks, magmatic rocks, and metamorphic rocks.
4. They are formed by the gradual solidification of the debris of rocks torn off by erosion.
5. They are formed by increasing temperature with depth and rocks from very high temperatures which form magmas.
6. They are formed at the depth of the earth due to high temperatures and high pressures which rise to its surface following geological phenomenons.
7. This is the study of stratified rocks.
8. This is the scientific study of fossils.
9. It is a prehistoric remain of living organisms.
10. This is the time a radioactive substance has its activity reduced by a factor of two.

Chapter 6: Questions

1. Explain the paleontological evidence of evolution.
2. What is relative dating?
3. What is absolute dating?
4. Name the two earth's crust.

Chapter 6: Answers

1. It is when the hard parts (the skeletons, the shells) of the sediments of the living organisms transform little by little to rock.
2. It is when we date the rocks in relation to each other.
3. This is when many igneous rocks form when the lava that cools contains radioactive elements.
4. Continental crust and oceanic crust.

Chapter 7: Questions

1. What is the smallest biological unit that can evolve over time?
2. What is anagenesis?
3. What is cladogenesis?
4. Define phylogeny.
5. What are clades?
6. What is biogeography?
7. What is gradualism?
8. What is diversity?
9. What is heredity?
10. What are alleles?

Chapter 7: Answers

1. Population.
2. This is the evolution of a specie without the splitting of the specie.
3. This is the evolution of a specie with the splitting of the specie.
4. This is the history of the evolutional development of an organism.
5. These are organisms that share a common ancestor.
6. This is the study of the geographical distribution of plants and animals.
7. This is a gradual change that occurs in species over a long period of time.
8. This is a variety in a specie.
9. This is the passing down of traits from one generation to another.
10. They are specific gene variations.

Chapter 8: Questions

1. What is a trait?
2. What are acquired traits?
3. Give an example of an acquired trait.
4. What are inherited traits?
5. Give an example of an inherited trait.
6. What is mutation?
7. What is a gene?
8. What is DNA?
9. What is a homologous structure?
10. What is an analogous structure?

Chapter 8: Answers

1. A trait is an inherited character.
2. These are learned traits.
3. Learning to write.
4. These are traits that we were born with.
5. The colour of the eyes.
6. This is a sudden change in DNA sequence.
7. A gene is a DNA fragment.
8. This is what contains the genetic information of living organisms.
9. It is a structure with similar organisation but different function.
10. It is a structure whose organization is different but its function is similar.

Chapter 9: Questions

1. Define microevolution.
2. Define macroevolution.
3. Give an example of microevolution.
4. Give an example of macroevolution.
5. What is convergent evolution?
6. What is divergent evolution?
7. What is the cause of convergent evolution?
8. What is the cause of divergent evolution?

Chapter 9: Answers

1. This is evolution on a large scale.
2. This is evolution on a small scale.
3. The changes in allele frequency.
4. Bird to reptile.
5. This is when non-related species become more similar.
6. This is when related species become more different from each other.
7. This is due to the fact that the non-related species adapt in the same environment.
8. This can be due to variations in biotic or abiotic factors.

Chapter 10: Questions

1. What is speciation?
2. What is allopatric speciation?
3. What is sympatric speciation?
4. What is the hybridization zone?
5. What are the conditions necessary for speciation?
6. What influences genetic variability?
7. What is genetic drift?
8. What can influence genetic drift?
9. What is the founder effect?
10. What is the bottleneck effect?

Chapter 10: Answers

1. Speciation is the appearance of new species during evolution.
2. This is when identical populations initially are separated geographically. They accumulate genetic differences over time and become two distinct species.
3. It is the formation of new species without geographical isolation.
4. It is the place of contact of two species.
5. Genetic variability, reproductive isolation, and natural selection.
6. Sexual reproduction, mutations, gene flow, genetic drift, and the transfer of horizontal genetic information.
7. It is the random change of the frequency of alleles in a small population.
8. The founder effect and the bottleneck effect.
9. It is when one or a few individuals migrate and start a new isolated population.
10. It is the decrease in population due to a sudden change in the environment such as natural disaster or human intervention.

Chapter 11: Questions

1. What are the barriers that promote reproductive isolation?
2. Name the prezygotic barriers.
3. What is geographic isolation?
4. Give an example of a geographic barrier.
5. What is ecological isolation?
6. What is behavioural isolation?
7. What is temporal isolation?
8. What is mechanical isolation?
9. Name the post-zygotic barriers.

Chapter 11: Answers

1. Prezygotic barriers and postzygotic barriers.
2. Prevention of gamete fusion, ecological isolation, geographic isolation, behavioural isolation, temporal isolation and mechanical isolation.
3. This is when species live in different areas.
4. A river.
5. This is when species live in the same area but different habitats, so they rarely meet.
6. Different species by their courtship display.
7. This is when species breed at different times of the day or in different seasons.
8. This is when structural differences between species prevent mating.
9. Natural selection, non-viable or sterile hybrids.

Conclusion

Thank you once again for downloading this book. I hope it has helped you gain more knowledge in evolution.

Please, if you enjoyed this book, I would like you to leave a review. It'd be appreciated.

Thank you.

www.ingramcontent.com/pod-product-compliance
Lightning Source LLC
Chambersburg PA
CBHW031510210526
45463CB00008B/3183